BLASTOFF!

MARS

BLASTOFF!

MARS

by Tanya Lee Stone

BENCHMARK BOOKS

MARSHALL CAVENDISH

NEW YORK

With special thanks to Roy A. Gallant, Southworth Planetarium, University of Southern Maine, for his careful review of the manuscript.

Benchmark Books
Marshall Cavendish Corporation
99 White Plains Road
Tarrytown, NY 10591-9001
www.marshallcavendish.com

Library of Congress Cataloging-in-Publication Data
Stone, Tanya Lee.
Mars / Tanya Lee Stone.
p. cm. — (Blastoff!)
Includes bibliographical references.
ISBN 0-7614-1233-6
1. Mars (Planet)—Juvenile literature. [1. Mars (Planet)] I. Title. II. Series.
QB641 .S76 2002   523.43—dc21   00-046775

Printed in Italy

1 3 5 6 4 2

Photo research by Anne Burns Images
Cover photo: JPL/Galaxy Picture Library

The photographs in this book are used by permission and through the courtesy of:
Photo Researchers: U.S. Geological Survey/Science Photo Library,7, 10, 43.
NASA/Science Photo Library: 15, 21, 36, 39, 41, 45. David A. Hardy/Science Photo
Library: 22. GE Astro Space/Science Photo Library: 24. Julian Baum/Science
Photo Library: 28. Science Photo Library: 35. Stock Montage/Superstock: 8.
SEF/Art Resource, NY: 12, Scala, Art Resource: 14. Astronomical Society of
the Pacific: 13, 56. JPL/NASA: 19, 26, 49, 51, 52. NASA/MSSS/Galaxy,31. Calvin
J.Hamilton/Galaxy 40. Malin Space Science Systems/ Galaxy: 47. NASA/USGS: 32.
Beagle 2/ESA/Galaxy, 54.

Book design by Clair Moritz-Magnesio

CONTENTS

1

DISCOVERY OF THE RED ONE

Why have human beings been attracted to Mars for thousands of years? What is it about the Red Planet that has captured our imagination more than any other body in the Solar System? Perhaps it is because Mars is the closest planet we can see. Maybe it is because there are interesting similarities between our planet and this fourth planet from the Sun. After all, Mars has polar caps, a weather system, seasons, and even a day that lasts nearly the same as an Earth day.

Since ancient times, human beings have been fascinated by Mars. The ancient Egyptians gave the planet its first recorded name—Har décher, the Red One. The Babylonians called it Nergal, Star of Death, and both the ancient Romans and Greeks named it after their god of war.

The ancient Greeks thought that Earth was the center of the Universe and that the Sun and other planets moved around Earth in circles. They believed that Mars was one of five traveling stars that revolved around Earth. But their theories did not accurately explain the actual motion of the "wandering stars," or planets, that people saw moving across the sky. For hundreds of years, astronomers grappled

This view of Mars is a composite of multiple images taken by the *Viking* orbiters from a distance of more than 1,550 miles (2,500 km). The pale streaks near the craters on the upper left are deposits of windblown dust.

with this problem and tried to come up with theories that explained what they saw in the heavens.

EARLY ASTRONOMERS TACKLE MARS

In 1576, before the invention of the telescope, the Danish astronomer Tycho Brahe set up the best observatory the world had ever known. He studied the heavens and calculated the movements of the planets with remarkable accuracy. He paid special attention to Mars and tracked the nightly change in its position among the stars throughout the year. His highly accurate position plots were far more important than Tycho could ever imagine. They enabled one of his assistants,

In 1609, Johannes Kepler published his Astonomia Nova (New Astonomy), *which outlined two of his three laws of planetary motion. It was his painstaking effort to calculate the orbit of Mars that helped lead to these important new theories.*

Johannes Kepler, to work out the actual motions of the planets. When Tycho died in 1601, he left his data to Kepler, who continued the work.

Kepler was a better mathematician than Tycho was. Tycho had asked Kepler to analyze the data he had collected about Mars. But unlike Tycho, Kepler supported the theory that the Solar System was heliocentric, that the Sun, not Earth, was the center of the Universe. Nicolaus Copernicus had first proposed the theory in a book published in 1543. Kepler used the heliocentric theory, as well as Tycho's data, to work out the laws of planetary motion. Prior to Kepler's laws, astronomers held a common idea that a planet moved around Earth in a perfect circle and always at the same speed.

Of all the planets, Mars was the most troublesome—it was rarely in the position in which it was predicted to be. Kepler boldly theorized that the idea of circular orbits was not correct. Using Tycho's data, he plotted the orbit of Mars in an elliptical, or oval-like, shape instead. It fit perfectly. Kepler's first law of planetary motion was announced in 1605—the planets orbit the Sun in ellipses.

He went on to form his second and third laws of planetary motion that had to do with how fast planets travel in their orbits and how long it takes for planets to revolve around the Sun. He also showed mathematically that the Sun did have a force that pulled on the planets and moved them along in their orbits. He was able to calculate the distance of Mars from the Sun at various points in that planet's orbit. Now more was known about Mars, but it would take the use of telescopes to be able to closely study the surface of the planet.

CHRISTIAAN HUYGENS AND GIOVANNI DOMENICO CASSINI

Christiaan Huygens grew up in a house that had a constant flow of intellectual visitors. His father, Constantijn, was a prominent Dutch thinker who embraced science as well as the arts. The young

The dark patch on the right is Syrtis Major, a volcanic shield probably composed of basalt, a type of rock. On the bottom is Mars's carbon dioxide ice cap, near the south pole.

Christiaan was exposed to such people as the poet John Donne, the painter Rembrandt van Rijn, and the philosopher René Descartes. In this rich environment, he became a great thinker and scientist. Christiaan loved to tinker with the newly invented telescope, improving lenses and making keen observations of the heavens. In fact, in his twenties, Huygens discovered Titan, Saturn's largest moon.

Huygens went on to make important discoveries about Mars. In 1659, he sketched, based on his observations, a dark area called Syrtis Major, which means "great wetland" in Latin. Scientists thought that Syrtis Major and other dark areas they saw on Mars were water—a main ingredient for the existence of life. In reality, this formation is simply a sloped area that has a bulge rising about 3.7 miles (6 km) from the planet's surface. Huygens also calculated the planet's period of rotation at about 24 hours. In 1666, an astronomer named Giovanni Domenico Cassini improved on Huygens's figure, claiming Mars completed a rotation once every 24 hours, 40 minutes. They were both extremely close. Mars's actual rotation is 24 hours, 37 minutes. Cassini also observed large white areas, which were polar caps, at the northern and southern poles of Mars. The caps, as well as other areas, would later be examined for the existence of water.

WILLIAM HERSCHEL

In the 1780s, astronomer William Herschel brought the world another step closer to understanding Mars. He measured the tilt of its axis— the imaginary line on which a planet rotates. The angle was very close to Earth's tilt, indicating that Mars could have seasons similar to our own planet. Through his telescope he also saw what he theorized was an atmosphere containing gases necessary to sustain life. Herschel not only believed that there could be intelligent life on Mars, he felt that this could be true for every planet or star, including the Sun.

CANALS AND MOONS

In 1877, a well-respected Italian astronomer named Giovanni Schiaparelli made a map of Mars based on his own observations. He included the dark and light patches that he had seen through his telescopes. He had also observed what he believed to be a pattern of lines crisscrossing the planet. He described these as *canali*, which means "channels" in Italian. However, it was incorrectly translated as canals, which are built by people. Although he was not the first person to comment on *canali*, the popular notion was that Schiaparelli had seen Martian-made canals. Other astronomers turned their telescopes to Mars as well. Their imaginations fueled by what Schiaparelli had described, many drew similar fictitious line patterns on their maps of Mars. This helped to further the belief that there was life on Mars.

The year 1877 was important for another reason. At the U.S. Naval Observatory in Washington, D.C., American astronomer Asaph Hall observed two satellites of Mars. The first one he found, on

In addition to investigating Mars, Giovanni Schiaparelli conducted extensive studies of comets. His theories and findings helped spark a growing interest in astronomy in the early 1900s.

August 11, was the outer satellite. It was named Deimos. Five days later, on August 16, he located the inner moon of Mars and called it Phobos.

A CONTROVERSIAL FIGURE BUILDS THE LOWELL OBSERVATORY

Percival Lowell was born in 1855 into a wealthy Boston, Massachusetts, family. He was interested in astronomy from the time he was very young and often carried a telescope to the roof of his house to look at Mars. He was a bright young man who went on to study at Harvard University, be named the foreign secretary to the Korean Special Mission to the United States, and become a millionaire

One of Percival Lowell's biographers wrote, "Of all the men through history who have posed questions and proposed answers about Mars, [he was] the most influential and by all odds the most controversial."

THE FIRST TELESCOPES

Telescopes began to be used in the early 1600s. Although there are some scientists who believe that there were earlier telescopes, Hans Lippershey is generally credited with building the first one in 1608.

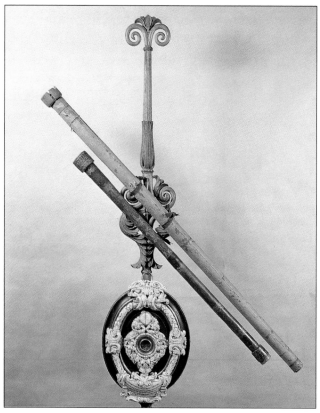

Upon hearing that a Dutch optician had arranged lenses to magnify distant objects, in 1600 Galileo crafted his own telescope and trained it on the night sky.

Galielo Galilei (as well as other craftsmen and scientists) quickly built one based on Lippershey's design and was able to magnify objects by three times. Galileo then improved on his design over the next year until he built a telescope that could magnify objects by twenty times. It was through this telescope that he made his first discoveries.

These early telescopes were very different from the complex instruments built today. They consisted of two tubes

Today telescopes have come a long way. The Hubble Space Telescope, for example, orbits Earth, providing images that are helping answer some of our questions about the Universe.

connected so that the length of the telescope tube could be adjusted to help bring an object into focus. At one end of the tube was a lens. The lens farthest from the viewer's eye was a curved convex lens. It was used to collect, and appear to bend, the light emitted from a distant object and bring it into focus. The second lens, placed at the eyepiece of the telescope, was a concave lens that straightened out the light, making it appear more normal to the viewer.

The design was later improved by swapping the positions of these lenses. Although it turned everything upside down to the viewer, the new design greatly improved the quality of the images.

Many of the earliest telescopes were 5 or 6 feet (1.6–1.8 m) long. But by the middle of the 1600s, telescopes had increased in length. In 1656, Christiaan Huygens made one that was 23 feet (7 m) long. As telescopes developed, magnification increased and the quality of the lenses improved. Just imagine what Galileo might have discovered if he had one of the many different types of telescopes that exist today—from the largest radio telescope dish in Arecibo, Puerto Rico, to the Hubble Space Telescope gathering data in space.

by the age of thirty. But when he was thirty-eight, he surprised all who knew him by announcing that he was giving up his career in business and politics in order to spend all his time studying the planet Mars. Lowell decided to build an observatory for that sole purpose. He chose Flagstaff, Arizona, for its low population (at the time) and clear skies. It opened on June 1, 1894.

Lowell was a controversial figure, viewed skeptically by the scientific world. Scientists wondered what he hoped to accomplish without any formal training in astronomy. To the general public, he was a wildly popular man. He appealed to large audiences with his observations and theories about irrigation canals and cities on Mars.

His ideas about intelligent life on the Red Planet were perpetuated by science-fiction accounts. It would take the hard evidence gathered by the instruments of the modern space age to dispel Lowell's dramatic theories.

Difficult to See

The progress in studying Mars was slow, considering that the technology required to study distant objects was also slow in developing. But remember that Mars is about half the diameter of Earth and it is more than 140 times farther from Earth than the Moon is!

2

A New Age of Exploration

Even before the first mission was sent to Mars, there was an attempt to contact the elusive Red Planet. In 1922 and 1924, during times when Mars was at its closest position to Earth, the U.S. government requested radio silence for specific periods of time so that astronomers would be able to detect any signals that might be coming from Mars. Naturally there were none. It would take the onset of the Space Age before scientists would mark some major new breakthroughs.

MARS I AND THE U.S. MARINER PROGRAM

The Space Age officially began on October 4, 1957, with the successful launch of a Russian satellite called *Sputnik I*. The Moon was the first target of interest, but attempts to reach Mars came soon after. The Russians launched *Mars 1* in November 1962. Soviet scientists stayed in contact with the probe until March 21, 1963. After that, radio transmissions ceased. Although the spacecraft passed Mars at a distance of 120,000 miles (193,000 km), it was no longer sending back data. Then the Soviet's next attempt did not even leave Earth's orbit.

One of the most successful missions to Mars was the
Pathfinder, *which landed on the Red Planet July 4, 1997.*

The next Mars mission was launched by the United States. The National Aeronautics and Space Administration (NASA) had already initiated its *Mariner* program. While *Mariner 1* had been destroyed, *Mariner 2* flew past Venus in 1962 at a distance of 20,900 miles (33,635 km). On November 5, 1964, *Mariner 3* was sent to Mars, but a malfunction cancelled any hopes that the mission would reach its intended target. The spacecraft is now in a solar orbit. Scientists scrambled to solve the problem before *Mariner 4*'s scheduled launch three weeks later. Their hard work paid off. On November 28, *Mariner 4* successfully took off for and later reached the Red Planet.

Compared to some of the spacecraft NASA creates today, *Mariner 4* was quite small. It was about 9 by 22 feet (2.7 by 6.7 m) and only weighed about 575 pounds (261 kg). Some thought it looked like a windmill hurtling through space. But this little wonder carried a television camera ready to show the world what it had waited to see for thousands of years—the surface of Mars.

It took 228 days for *Mariner 4* to get to Mars. The craft arrived on July 14, 1965. The twenty-two images that were transmitted to Earth were a huge disappointment to some, a wondrous sight to others. People saw a dry, dusty surface riddled with craters. There was no water in sight. From this glimpse of Mars and the information *Mariner 4* gathered about the planet's thin atmosphere, it did not seem at all probable that Mars could sustain life. Still, there was no conclusive evidence one way or the other.

After *Mariner 4*, the Soviets made an unsuccessful attempt at Mars with their *Zond* program. The United States launched *Mariners 6* and 7 and collected 201 photographs. The Soviet Union then sent *Mars 2* and 3. But the next real breakthrough came with *Mariner 9* (*Mariner 8* had crashed). It was launched on May 30, 1971, and reached Mars 167 days later. It was the first spacecraft to orbit another planet. The instruments aboard *Mariner 9* had been improved since the earlier *Mariners*. *Mariner 9* transmitted more than 7,300 images from all over the planet over the course of a year. It also took twenty-seven photographs of Mars's moons Phobos and Deimos. In contrast, *Mariner 4* had been able to cover only a small percentage of Mars.

An intricate landscape was unveiled to the world by *Mariner 9*. It included enormous volcanoes, a canyon nearly 3,000 miles (4,828 km) long, frozen underground water in the form of permafrost, and what appeared to be dried-up riverbeds. "The Mars revealed by *Mariner 9* was not one-dimensional; it was an intriguingly varied planet with a mysterious history. The possibility of early life once more emerged," said planetary scientist Bruce Murray.

The missions of the 1970s added to our ever-growing knowledge of Mars's varied terrain. This image, taken from 350,000 miles (563,000 km) away, shows the giant volcano Olympus Mons (top) and the several volcanoes that make up the Tharsis Bulge (center).

This illustration shows the orbital paths of Viking I *(purple) and* Mariner 9 *(blue) around Mars. The red lines indicate the orbits of Mars's two moons—Deimos and Phobos.*

VIKING 1 AND 2

After the success of *Mariner 9*, NASA was more determined than ever to press on. The next goal was to land a craft on the surface of Mars and begin the search for any signs of living organisms. While the Russian Mars program continued to make attempts that met with failure, the United States was planning *Viking 1* and *Viking 2*, aimed at two different sites on the planet.

Each *Viking* was equipped with an orbiter coupled with a lander. The lander would separate from the orbiter and gently come to rest on Mars. *Viking 1* was launched in August 1975 and *Viking 2* was close behind, sent a month later. On July 20, 1976, came the historic announcement. "Touchdown! We have touchdown!" cried flight controller Richard Bender. *Viking 1* had landed on its target site and was intact. *Viking 2* followed, landing successfully on September 3.

For several years, *Viking* gathered a wealth of new information about the surface conditions on Mars. The laboratories on the landers were designed to function 79 million miles (127 million km) from Earth and, incredibly, were doing just that! Information about soil conditions, the seasons, temperature ranges, wind speeds, and cycles of day and night was recorded. But the real excitement centered around three experiments that sought to detect life on the Red Planet. One experiment searched for any evidence of an organism that ingested food, another for any trace of organism that breathed. The third experiment was designed to determine if any organisms containing organic materials had died and left their remains on Mars.

Although some positive results generated some initial excitement, scientists soon discovered there were serious flaws in the experiments. Ultimately, they did not find any signs of life. But together, the two *Vikings* had collected more than 50,000 images of Mars and mapped 97 percent of the planet. When the second *Viking* had its last moments

An artist's rendering of the Mars Observer. *One of the many goals of the mission was to study dried-up riverbeds, dust, fog, and frost on the planet's surface. But much to the dismay of scientists, the spacecraft malfunctioned, and NASA had to try again.*

of contact, project manager George Gianopulos said of the spacecraft, "It's like losing an old friend, how do you express it?"

FAILED MISSIONS

Following *Viking*, there were several U.S. and Russian missions that failed. In July 1988, Russia sent *Phobos 1* and *2* into space. Some data were gathered, but scientists lost contact with *Phobos 1* a month later and with *Phobos 2* in March 1989. In September 1992, the U.S. *Mars Observer* was launched. But when NASA sent a signal to the *Observer* in August 1993, there was no response. Nobody has ever figured out what happened to this spacecraft. Another attempt at Mars was made by Russia in November 1996 when that nation launched *Mars 96*. Again disaster struck when *Mars 96* failed to break free of Earth's gravity and eventually burned up.

Scientists experienced continued bad luck in successfully reaching Mars, and there were two more U.S. failures to come. The *Mars Climate Orbiter* was launched in December 1998 and was lost nine months later. The *Mars Polar Lander* was launched in January 1999. But as the world watched in anticipation and waited for the *Polar Lander* signal to come back on December 3, 1999, nothing happened. Time passed and NASA grappled with reasons to explain its fate, but the *Polar Lander* had disappeared. Another failure.

MARS GLOBAL SURVEYOR AND MARS PATHFINDER

Despite these setbacks, there had been success a few years earlier with two U.S. space probes that followed in *Viking's* path and reached the Red Planet. On November 7, 1996, NASA launched the *Mars Global Surveyor*. The *Pathfinder* took to the sky on December 4, 1996. The *Pathfinder* got there first, traveling a different path than the *Surveyor*. This time, scientists had taken a new approach to landing on Mars.

Instead of worrying about the craft being destroyed on impact, it was designed to be cushioned by massive air bags. It bounced more than fifteen times before coming to a stop on America's Independence Day—July 4, 1997.

The next day, a vehicle called the Sojourner rover rolled down a ramp and out of the *Pathfinder* lander. At only 26 inches (66 cm) long, 18 inches (46 cm) wide, and less than 12 inches (30 cm) high, the com-

In July 1997, Sojourner inched its way across the surface of the Red Planet. The data collected by this small rover helped astronomers classify three types of Martian rocks.

pact Sojourner had an enormous job to do. Controlled by a computer, it was able to travel around large obstacles, climb over small ones, and inch its way across the Martian landscape collecting rocks to test and transmitting images.

The *Pathfinder* mission gathered an impressive array of information about Mars. Images of a dusty, flat landscape riddled with rocks were sent back to Earth. In addition, several dust devils were located and measured for temperature, wind speed, and pressure. These compact dust storms are at least partially responsible for sending dust into Mars's atmosphere. There it absorbs most of the solar radiation the planet receives.

The mission also gathered evidence that would seem to support the notion that there was once running water on the Red Planet. First, the distribution of different sizes of rock are in keeping with what scientists believe would have happened during a possible ancient flood. Also smooth, rounded rocks and pebbles normally formed by water running over them were found on the planet's surface.

Scientists thought at first that the Sojourner would last about a week and the *Pathfinder* about a month. But the *Pathfinder* continued to transmit data until September 27, 1997 and the Sojourner until October 8, 1997. It was an extremely successful mission that collected more than 16,000 images, provided us with a better understanding of the Martian surface, and took more accurate measurements of the atmosphere than the *Vikings* had.

"I want to thank the many talented men and women at NASA for making the mission such a phenomenal success. It embodies the spirit of NASA, and serves as a model for future missions that are faster, better and cheaper. Today, NASA's *Pathfinder* team should take a bow, because America is giving them a standing ovation for a stellar performance," said NASA administrator Daniel S. Goldin.

Meanwhile, the *Mars Global Surveyor* got to work. The *MGS* was an orbiter, so it was not going to land on the surface of Mars. Thus, it

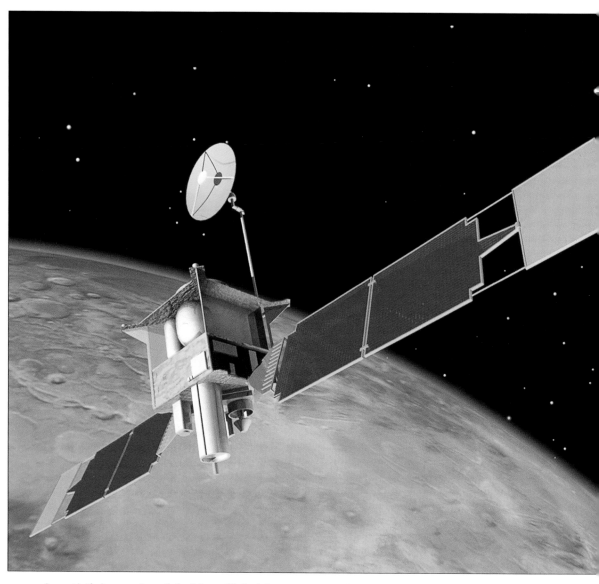

An artist's impression of the Mars Global Surveyor *in orbit. Fulfilling the objectives of the failed Mars* Observer *mission, the MGS has already given scientists more than three years of data to analyze.*

was designed to last much longer than the *Pathfinder*, with an expected life span of one Martian year—687 days. The *MGS* arrived at Mars on September 11, 1997. One of the first things it discovered was that Mars had traces of a magnetic field. This was one piece of information that had eluded scientists for a long time and was related to whether Mars had ever sustained life.

The *MGS* continues to gather and transmit data and will remain active as long as possible. It orbits the planet about twelve times a day and has already been going strong for more than three years. The images that it records are up to fifty times clearer than previous technology had allowed. These enhanced images give scientists a much more accurate view of the planet than they have had in the past. Much of the detailed knowledge we have of environmental and geographical features of Mars is due to the success of the *Mars Global Surveyor*.

Getting There

If you were driving 60 miles per hour (96.5 km/hr) in a car, it would take 271 years, 221 days to get to Mars from Earth.

3

THE MARTIAN ENVIRONMENT

We have learned a great deal about Mars from successful space missions. Scientists have interpreted data and images transmitted from both orbiters and surface landers to give us a greater understanding of this fourth planet from the Sun.

EARTHLIKE PATTERNS

Scientists have made a few comparisons between Earth and Mars. Because Mars makes a complete rotation once every 24 hours, 37 minutes, the length of a day on Mars is similar to that on Earth. And although it is much colder on Mars than on Earth, Mars also experiences seasons. This is because both planets have a similar axis tilt—Earth is on a 23.45-degree tilt while the tilt of Mars's axis is 25.19 degrees.

The *Viking* mission recorded Earthlike seasonal patterns on Mars. There are even times when temperatures can reach more than 70 degrees Fahrenheit (21° C) at the Red Planet's equator. But the main difference between the seasonal patterns on Mars and Earth is that Mars's seasons last about twice as long as Earth's. This is because Mars revolves around the Sun in approximately two Earth years.

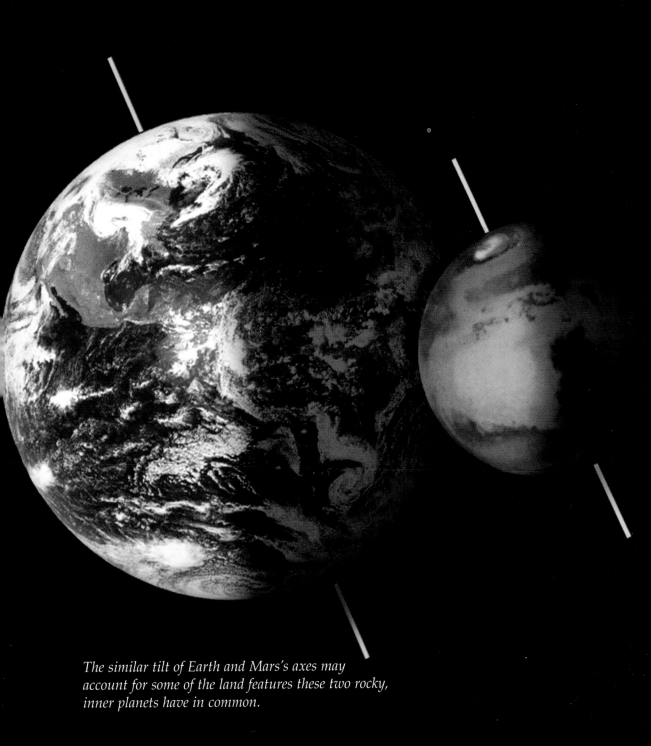

The similar tilt of Earth and Mars's axes may account for some of the land features these two rocky, inner planets have in common.

THE ATMOSPHERE AND CLIMATE

Carbon dioxide was the first gas known to be present in the Martian atmosphere. Astronomer Gerard Kuiper initially proposed his theory of this in 1947. A few years later, scientists surmised that the atmosphere was made up of a range of additional gases. The *Viking* mission then took detailed measurements of the Martian atmosphere, confirming that it is more than 95 percent carbon dioxide. The findings of the mission also confirmed the presence of small amounts of oxygen, nitrogen, and argon.

Clouds were detected as early as 1877, and we now know that Mars has a very thin cloud cover. The atmosphere keeps temperatures

The white area at the top of this Viking *image shows a Martian polar ice cap.*

on Mars generally well below freezing, as it does not provide adequate insulation of heat. In the winter, almost 20 percent of the air freezes. Much of the resulting ice remains locked up in the polar caps until spring warming turns it back into gas. Although the average temperature on Mars is minus 67 degrees Fahrenheit (-55° C), the temperature can range from minus 205 degrees to 72 degrees Fahrenheit (-132–22 ° C) in the summer. But these higher temperatures rarely reach more than a few feet above the soil.

We first learned about the Martian climate from data collected by *Mariner* and *Viking*. But later information gathered by the Hubble Space Telescope and the National Radio Astronomy Observatory telescope provided new evidence of abrupt climate changes that occur on Mars. In part, Mars's elliptical orbit causes these changes.

When Mars is at aphelion, or farthest from the Sun, the colder conditions help to form water ice clouds in the atmosphere. These clouds form around surface dust, which remains at lower altitudes during aphelion. This reduces temperatures further, causing the dust to freeze and cover the ground.

In contrast, when Mars is at perihelion, or closest to the Sun, it gets 40 percent more sunlight than during aphelion. The increased temperatures during perihelion contribute to the huge dust storms that are frequent at this time. The storms send the dust miles into the air where it absorbs sunlight, raising the atmospheric temperature even further.

A SALMON SKY

The soil on Mars is iron-rich clay, which gives it a rusty appearance. Winds that reach speeds of nearly 125 miles per hour (201 km/hr) swirl the dirt in violent dust storms that can hit at any time. When the storms carry dust into the Martian atmosphere, the sky takes on a salmon pink glow. These storms continually change the appearance of

PHOBOS AND DEIMOS:
THE TWO MOONS OF MARS

Mars's two moons—Phobos and Deimos—are small and oddly shaped. Because they reflect so little light, Phobos and Deimos are among the darkest objects in the Solar System.

Mars's two moons are actually asteroids that were pulled in and captured by the planet's gravitational field. Phobos's orbit is 5,814 miles (9,378 km) from Mars. At Phobos's widest point, it is only about 13 miles (21 km) across. The Stickney Crater on Phobos, named after discoverer Asaph Hall's wife Angelina Stickney, is 6 miles (9.6 km) wide. Phobos orbits Mars every 7 hours, 39 minutes. Twice during the Martian day, Phobos rises in the west and sets in the east. Because its orbit is moving ever closer to the planet, this little satellite will ultimately face a violent end. In about 50 million years, it will either crash into Mars or be broken apart by the planet's gravity.

Deimos is even smaller than Phobos. At about 8 miles (12 km) wide, it is one of the smallest satellites in the Solar System. At a distance of 14,545 miles (23,459 km), it orbits Mars every 30 hour, 18 minutes. Both moons are made of rock and ice.

Some scientists believe that future missions to Mars

might include using Deimos as a landing base. From the surface of that satellite, astronauts could send out probes and rovers to the Martian surface. The astronauts could then analyze samples without having to bring them back to Earth.

This computer composite image shows the pockmarked surface of Phobos. Some astronomers have compared Phobos and its twin satellite, Deimos, to potatoes.

The rusty appearance of the rock-strewn surface of Mars. Winds whip across the planet, sometimes kicking up tremendous dust storms. The dust settles only to be blown about again.

the Martian surface as the dust resettles each time. Every few years, the dust storms grow so large that they envelop the entire planet. In fact, *Mariner 9* had to wait several weeks for a dust storm to clear before continuing on its mission.

In February 2000, the *Mars Global Surveyor* cameras witnessed one of Mars's tornado-like dust storms for the first time. The *MGS* captured images of the wind whirling across the surface creating lines in the sand. We now know that these storms account for at least some of the veinlike tracks that have long been observed in the Martian sand.

Gravity

The gravitational field of Mars is a little more than a third that of Earth's. This means that a 100-pound (45.4-kg) person on Earth would only weigh 38 pounds (17.2 kg) on Mars.

4

THE MARTIAN TERRAIN

Millions of measurements that the *Mars Global Surveyor* has taken have resulted in the first three-dimensional map of Mars. From these new maps, we can view elevation levels and topographical features. We knew from previous missions that Mars is a cold, dusty, rocky place. But the three-dimensional maps compiled from *MGS* information have given scientists valuable new views of the planet.

We now know, for example, that in the southern hemisphere the terrain is heavily cratered and the elevation is much higher than in the northern hemisphere. In contrast, the northern hemisphere consists mainly of lowland—areas of lower elevation with fewer craters. One exception is the Tharsis Bulge, which was created by volcanic activity in the early history of the planet.

Mars was formed about 4.5 billion years ago. Scientists have theorized that it has a metallic core surrounded by a mantle of molten rock and then a thin rock crust. In its early history, Mars was probably warm. The internal heat of this ancient Mars was responsible for the volcanic areas on the planet. It is their inactive remains that we see today.

This three-dimensional map of Mars is color-coded to show varying elevations. Yellow areas are of average elevation, with red indicating higher areas and blue and green marking the lowlands. Up to 27 million measurements were made in 1998 and 1999 by timing laser pulses sent to the planet's surface.

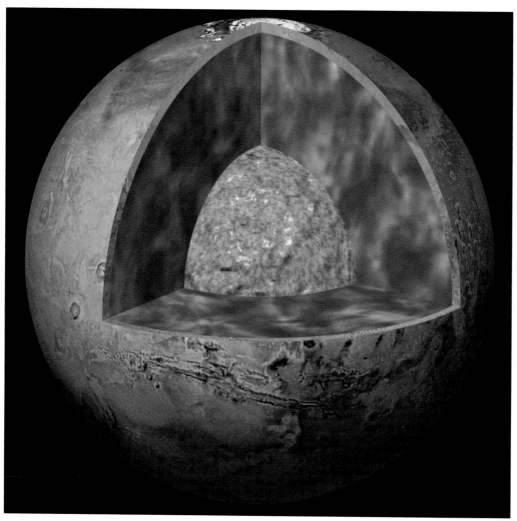

Scientists will continue to study the surface of Mars, but they want to know more of its interior as well. The planet may have a molten metal core surrounded by a molten rock mantle, topped by a thin rock crust.

POLAR CAPS

Mars has two polar caps—one on the northern end and one on the southern end of the planet. Information gathered by the *Viking* mission indicated that the southern polar cap is made up of frozen carbon dioxide, while the northern cap is water ice. *Viking* observed much less water vapor rising from the southern cap than the northern cap during the warmer seasons. However, in the spring and summer, the southern polar cap shrinks dramatically.

    MGS pictures taken in the year 2000 indicate that the northern and southern caps also have a distinct difference in appearance. Although earlier missions showed that the polar cap surfaces varied

The Mars Global Surveyor *sent some 2.6 million laser pulses to help compile this image of Mars's north pole. Made of water ice, the polar cap is about 745 miles (1,200 km) across and 2 miles (3.2 km) thick.*

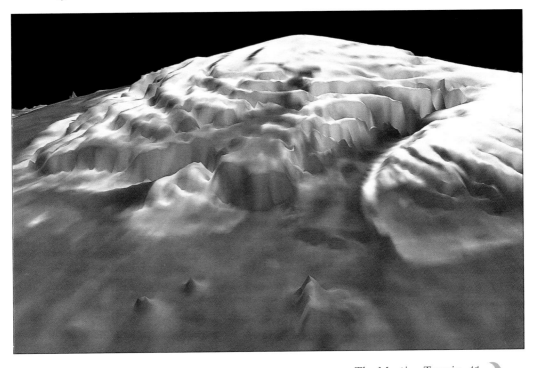

from one another, *MGS* pictures now illustrate this in much greater detail. From this new information, scientists have given nicknames to the textures found at each pole—the northern polar cap is described as cottage cheese and the southern as Swiss cheese.

VOLCANOES, CANYONS, PLAINS, AND SAND DUNES

Mars sports a varied terrain. One of the amazing discoveries of the *Mariner 9* mission was that Mars has enormous volcanic mountains. In fact, Mars is home to the highest peak in the Solar System—Olympus Mons. At 15 miles (24.1 km) tall, it stands three times higher than the highest mountain on Earth—Mount Everest. But venturing up Olympus Mons would not be nearly as dramatic as reaching the peak of Mount Everest. The Martian volcano climbs so gradually that you might not even realize you were on the highest mountain in the Solar System. It is called a shield volcano because of its wide base. With a diameter of nearly 375 miles (603 km), Olympus Mons covers an area the size of Arizona.

There are other vast volcanic mountains on Mars as well. The Tharsis Bulge has three large volcanoes that form a chain. Mars's inactive volcanoes range in age from 20 million to 2.5 billion years old. Although no Martian volcanoes are active, areas called hot spots have been found. Hot spots are places where the temperature of the ground is higher than on the rest of the surface. These spots may be from the heat released from radioactive elements in the soil.

Mariner 9 also found a canyon system on Mars that makes Earth's Grand Canyon seem tiny. It was named the Valles Marineris, or the Grand Canyon of Mars. It is 3,000 miles (4,828 km) long—more than the width of the United States. The Valles Marineris is 400 miles (644 km) wide, which is 20 times wider than the Grand Canyon.

There are also expansive pink sand dunes on Mars's surface. In some areas, the landscape looks very much like a desert on Earth. In

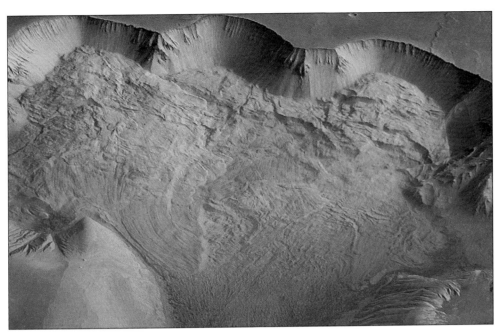

The Ophir Chasm is located in the central region of the Valles Marineris, the huge canyon that prominently marks Mars's surface.

fact, the Solar System's largest dune field is located at the northern polar cap on Mars. The dunes encircle the cap, and *MGS* images have shown that the sand and snow sometimes mix. This indicates that the dunes of Mars are active.

Mars is also home to Hellas, a vast featureless plain that covers an area as large as the Caribbean Sea. At 1,300 miles (2,092 km) wide and 6 miles (9.6 km) deep, this enormous basin was created by the impact of asteroids crashing into the surface of Mars. Scientists believe that Hellas was formed about 4 billion years ago.

METEORITES FROM MARS

Meteorites are pieces of rock and metal that fall to Earth. They have been found all over the world. More than a dozen meteorites are known to have come from Mars. Scientists can tell where they came from because they hold trapped gases that match those in the atmosphere of Mars. More evidence that there was once water on Mars has also been found in some of these meteorites. Several of them contain water-soluble chlorine, indicating that they were once in contact with seawater.

The most famous meteorite from Mars is ALH84001. In 1984, this meteorite was discovered in Antarctica. It fell from Mars about 13,000 years ago. The meteorite was dubbed ALH84001 because it was found in the Allan Hills in 1984 and it was the first one found that year.

ALH84001 is estimated to be about 4.5 billion years old, which is about the time that Mars was formed. The meteorite contains microscopic tubelike formations that some scientists think may be microfossils of bacteria. Carbonates, which are found in bones, have also been detected in ALH84001. Some scientists believe that this could be evidence of ancient life on Mars. Others disagree, stating that these clues could simply be bacteria that arose while testing and preparing the sample or that they could be bacteria from Earth that managed to get inside crevices of ALH84001. The debate will no doubt continue until proof can be found to support the theory of life having once existed on Mars.

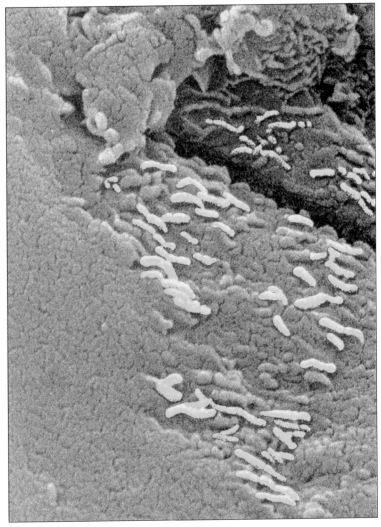

Proof that life once existed on Mars? These grain-shaped structures, colored yellow here, were found in the Martian meteorite ALH84001. They may be microfossils of bacteria-like organisms that lived on Mars billions of years ago. But some scientists doubt it.

The Size of Mars

Mars is a little more than half the size of Earth, with a diameter of 4,223 miles (6,796 km).

Around the Sun

Mars revolves around the Sun once every 687 days, or about every 23 Earth months. It is approximately 142 million miles (228 million km) from the Sun.

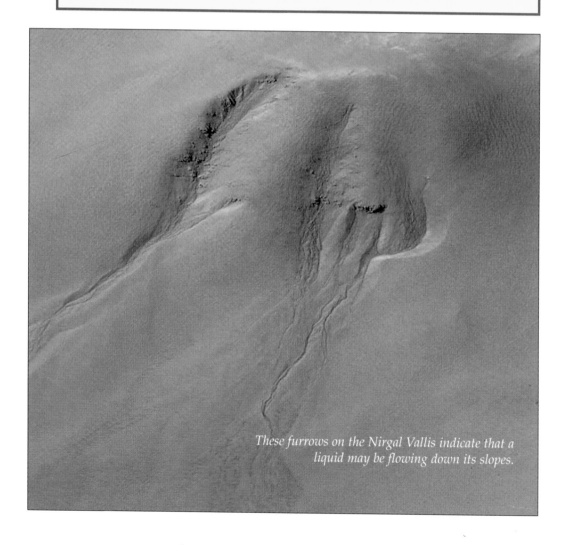

These furrows on the Nirgal Vallis indicate that a liquid may be flowing down its slopes.

ANCIENT OCEANS AND RIVERBEDS?

Astronomers have recorded channels on Mars that appear to be dried-up riverbeds. In fact, *Mariner 9* recorded thousands of them, both large and small. Water is necessary to sustain life. Therefore, the search for evidence that there was once water on the planet is a major focus of current Mars research. Scientists believe that running water formed the channels. While some of them may be the remains of ancient riverbeds, others appear as though they were carved during a major flood that may have occurred in Mars's past.

There is also evidence of an ancient ocean on Mars. The three-dimensional map that the *MGS* has created shows an enormous flat surface. At the edge is an area that resembles a shoreline with a series of stepped terraces. Shorelines on Earth display similar features—horizontal lines left on the surface at points where the shoreline gradually receded.

Although the frigid climate on Mars has seemed to rule out the possibility of liquid water flowing on its surface today, it is almost certain that there was once water there. Michael Malin, one of the people in charge of the imaging systems for the *MGS*, says, "There is ample evidence that a fluid with properties very much like water once flowed on Mars."

In June 2000, Michael Malin and Kenneth Edgett displayed *MGS* images that may indicate the presence of liquid water at the surface. These images show that something is flowing down the slopes of the Nirgal Vallis—a 250-mile-long (402-km) valley—in the same way that water would run through sand. Many scientists believe this is strong evidence that, given the right climate conditions, there could be groundwater reaching the Martian surface in certain locations.

5

FUTURE MISSIONS TO MARS

Although there have been more unsuccessful attempts to get to Mars than successful ones, every mission is important. Critical information is learned from our mistakes, and improvements are continually being made to the space program. The same is true for developing technology. Even though it might be lost in space, each piece of equipment that is conceived and created by scientists gets us that much closer to our goals. Of course, the ultimate goal is for a mission to be successful and for the expensive and innovative technology to be put to the test where it will be most useful—Mars.

Raising funds to pay for future missions is always a challenge, and plans are often changed. However, there are several missions that are scheduled to take place at the beginning of the twenty-first century.

MARS SURVEYOR 2001

On April 7, 2001, the *Mars Surveyor Orbiter* is scheduled to launch. If everything goes as planned, it should arrive in October 2001. The *Orbiter* will take three main instruments to Mars in order to conduct

Mars Surveyor 2001 Orbiter
Science Orbit Configuration - GRS Boom Deployed

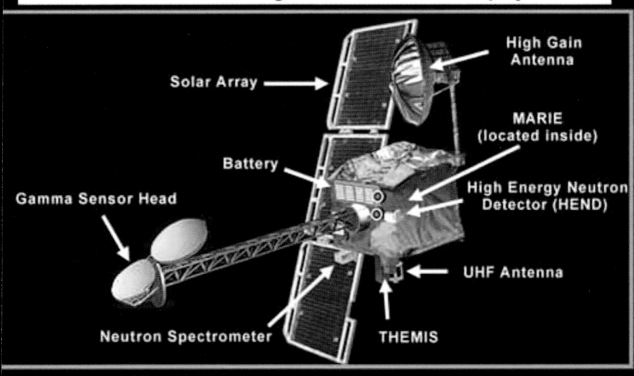

High Gain Antenna

Solar Array

MARIE (located inside)

Battery

High Energy Neutron Detector (HEND)

Gamma Sensor Head

UHF Antenna

Neutron Spectrometer

THEMIS

The Mars Surveyor Orbiter *2001 mission is aimed not only to broaden our understanding of Mars, but to help us understand the processes through which Earth's climate changes as well.*

experiments. The first is the Thermal Emission Imaging System (THEMIS). Using a powerful camera and an instrument called a spectrometer that measures the wavelengths of light, THEMIS will take up to 15,000 images in order to map the Martian surface. The second is the Mars Radiation Environment Experiment (MARIE). This instrument will evaluate the amount of radiation in the environment and measure how much radiation would be absorbed by human tissue in the event people are ever sent to explore Mars.

The third *Orbiter* instrument is the Gamma Ray Spectrometer (GRS). A previous GRS was lost aboard the Mars *Observer,* and it has been rebuilt for this new mission. The GRS has four main goals. The first is to determine how thick the polar caps are and how they change over time. The second is to map water that may be near the surface. The GRS will also determine the percentage of elements—especially hydrogen—that make up the surface, including the polar caps. Finally, it will study the cosmic gamma-ray bursts that reach Mars.

THE MARS 2003 ROVER MISSION

Because of the failed Mars missions in the late 1990s, the *Mars Surveyor* 2003 and *Mars Surveyor* 2005 missions are both under review by NASA and their scheduled launches may be delayed. However, in August 2000, NASA announced that it would first launch two identical 300-pound (136-kg) rovers to explore Mars. The rovers will both be launched on rockets from Cape Canaveral, Florida. The first is planned for May 22, 2003, and the second will follow on June 4, 2003. Information continuing to come in from *MGS* will help scientists choose the rovers' final destinations.

The rovers will travel to different locations, sites that will probably be selected as potential places to detect water. They will take panoramic images with a special digital camera and set out to explore each day. Each rover will be equipped with instruments that will analyze soil and rock samples. The rovers will ultimately be in search of evidence of life on the Red Planet.

THE ESA'S FIRST ATTEMPT TO EXPLORE MARS

The European Space Agency (ESA) has planned its first voyage to Mars in 2003 with the *Mars Express* mission. The mission is a cooperative effort among hundreds of international scientists. It will be launched from Russia and should reach Mars by late 2003 or early

Mars Surveyor 2003 and Mars Surveyor 2005 will both carry identical rovers to explore the surface of the Red Planet.

2004. The plan is for the orbiter to operate for at least one Martian year. But the craft has been designed to last twice as long.

One of the main goals of this mission is to find water on the Red Planet. Seven scientific instruments aboard the orbiter will conduct experiments to learn more about the atmosphere, surface, and struc-

One main goal of future missions is to locate water on Mars. Scientists will learn much more about the planet's surface in the process.

ture of Mars as well. Within the first Martian year, the orbiter should have the opportunity to gather data from the entire planet's surface.

Included in the orbiter's instruments is the High Resolution Stereo Camera (HRSC), with which images of the surface will be taken. An IR Mapping Spectrometer will use infrared technology, which allows us to "see" beyond the visible light spectrum, to look at Martian rocks and soil. Measurements of the Martian interior will also be taken with a device called the Radio Science Experiment (RSE).

Other instruments will be used to study the atmosphere. The UV Atmospheric Spectrometer will use ultraviolet technology to take measurements of the atmosphere's makeup. A Planetary Fourier Spectrometer (PFS) will use infrared technology to chart the temperature and pressure of the Martian atmosphere. The relationship between the solar wind and the upper atmosphere will be looked at with the Energetic Neutral Atoms Analyzer. The solar wind is a stream of charged particles given off by the Sun that travels throughout the Solar System.

In addition to these instruments, a lander called Beagle 2 will be aboard the mother spacecraft. Named after the ship the scientist

Students Take FIDO For a Ride

Some lucky high school students in New York, Alabama, Missouri, and Arizona got to participate in test runs with a model rover. The Field Integrated Design and Operations (FIDO) rover was built to practice communicating with and operating the rover. Students learned to command the rover and helped scientists with the field tests.

An artist's impression of Beagle 2 after it has successfully landed on the surface of Mars.

Charles Darwin sailed, the Beagle 2 will not only land on the Martian surface—it will bore beneath it. Since the atmosphere on Mars does not support life on the surface, the Beagle will use a drill and grinder to study rocks and soil below the surface. Armed with a small rover, cameras, a microscope, and other instruments, the Beagle 2 will be able to tell what types of rocks it finds, what the makeup of the rocks is, and if there are any fossilized bacteria inside them. The mission of Beagle 2 is to search for any signs of past or present life on Mars.

NASA LOOKS AHEAD

In October 2000, NASA announced new plans for future missions to Mars. The ambitious projects scheduled take us all the way to the year 2016. The first is the launching of a scientific orbiter to Mars in 2005. Called the *Mars Reconnaissance Orbiter*, the main goal of this trip will be to analyze the surface in such detail that rocks as small as 8 to 12 inches (20.3–30.4 cm) long will be studied.

In 2007, NASA hopes to be ready to launch a scientific laboratory that can function in space for long periods of time. This laboratory would be a prime candidate for sample-return missions to Mars, which have been long awaited. The hope is that such a laboratory would be able to land at multiple sites and collect samples of the Martian surface, eventually delivering them back to Earth for analysis.

Still other ideas for sample-return missions are being explored. Some could be ready to go by 2011. More conservative estimates would have a spacecraft on the launch pad by 2014, with a second to follow in 2016.

The role that NASA, as well as international space agencies, plays is crucial as it prepares for some of the most technologically demanding work ever pursued. What is essential to NASA's success is flexibility. As Scott Hubbard, NASA's Mars program director stated, "We

Once shrouded in mystery, steadily the Red Planet is giving up its secrets. With so many missions planned for the future, Mars will become the Solar System's high-traffic zone. But the benefits of this intense scrutiny are boundless.

have developed a campaign to explore Mars unparalleled in the history of space exploration. It will change and adapt over time in response to what we find with each mission. It's meant to be a robust, flexible, long-term program that will give us the highest chances for success." Whatever the future holds, there is no doubt that what we know now about the Red Planet is a mere fraction of what is yet to be revealed.

Mars Fact Sheet

Mean Distance from Sun: 142 million miles (228 million k)
Diameter: 4,217 miles (6,787 km)
Mean surface temperature: -45.67° F (-43.15° C)
Surface gravity: .38
Period of revolution (year): 686.98
Period of rotation (day): 1.026
Number of satellites: 2, Phobos and Deimos

GLOSSARY

aphelion The farthest point from the sun in an object's orbit around it.

astronomy The study of the planets, stars, and other heavenly bodies.

axis The imaginary line on which a planet spins.

concave Curved inward.

convex Curved outward.

meteor Matter from space that enters the Earth's atmosphere.

meteorite A meteor that has fallen to the Earth's surface.

opposition The time when a planet farther from the Sun than Earth is closest to Earth and opposite the Sun.

orbit The path of a heavenly body as it revolves around another heavenly body.

perihelion The nearest point from the sun in an object's orbit around it.

rotation The process of turning on an axis; or, one complete turn around an axis.

FIND OUT MORE

BOOKS FOR YOUNG READERS

Bortz, Fred. *Martian Fossils on Earth?* Brookfield, CT: Millbrook Press, 1997.

Bredeson, Carmen. *Our Space Program.* Brookfield, CT: Millbrook Press, 1999.

Clay, Rebecca. *Space Travel and Exploration.* New York: Twenty-First Century Books, 1996.

Cole, Michael D. *Living on Mars: Mission to the Red Planet.* Springfield, NJ: Enslow Publishers, 1999.

Couper, Heather, and Henbest, Nigel. *Is Anybody Out There?* New York: Dorling Kindersley, 1998.

Fradin, Dennis Brindell. *Is There Life on Mars?* New York: Simon & Schuster, 1999.

———.*The Planet Hunters: The Search for Other Worlds.* New York: Simon & Schuster, 1997.

Gallagher, Amie. *Outer Space: The Inner Planets.* Danbury, CT: Grolier Educational, 1998.

Getz, David. *Life on Mars.* New York: Henry Holt & Company, 1997.

Kelch, Joseph W. *Millions of Miles to Mars: A Journey to the Red Planet.* Parsippany, NJ: Silver Burdett Press, 1995.

Ride, Sally, and O'Shaughnessy, Tam. *The Mystery of Mars.* New York: Crown, 1999.

Vogt, Gregory. Scientific American Sourcebooks: *The Solar System.* New York: Twenty-First Century Books, 1995.

Other Books

Clarke, Arthur C. *The Snows of Olympus: A Garden on Mars*. New York: W.W. Norton, 1995.

Goldsmith, Donald. *The Hunt for Life on Mars*. New York: Dutton, 1997.

Henbest, Nigel. *The Planets: A Guided Tour of Our Solar System through the Eyes of America's Space Probes*. New York: Viking, 1992.

Pasachoff, Jay M., and Menzel, Donald H. *A Field Guide to the Stars and Planets*. New York: Houghton Mifflin, 1992.

Raeburn, Paul. *Mars: Uncovering the Secrets of the Red Planet*. Washington, D.C.: National Geographic Society, 1998.

Sagan, Carl. *Cosmos*. New York: Random House, 1980.

———. *Pale Blue Dot: A Vision of the Human Future in Space*. New York: Random House, 1994.

Sheehan, William. *The Planet Mars*. Tucson, AZ: The University of Arizona Press, 1996.

———.*Worlds in the Sky: Planetary Discovery from Earliest Times through Voyager and Magellan*. Tucson, AZ: The University of Arizona Press, 1992.

Wilford, John Noble. *Mars Beckons*. New York: Knopf, 1990.

Wunsch, Susi Trautmann. *The Adventures of Sojourner: The Mission to Mars that Thrilled the World*. New York: Mikaya Press, 1998.

WEBSITES

To visit NASA's Mars Exploration Program home page go to: http://mpfwww.jpl.nasa.gov/

To visit NASA's Ames Space Science Division Center for Mars Exploration, go to: http://cmex-www.arc.nasa.gov/

To visit NASA's Mars Team Online home page, go to: http://quest.arc.nasa.gov/mars/ This is an online project for kids to participate in and get updated mission news, as well as offer opportunities to chat with NASA experts online, to learn about background material on Mars, and many other activities.

National Geographic's Destination: Mars website gives kids information on upcoming missions to Mars, including plans to send humans to the Red Planet. There are also links to other good websites here. Visit: http://www.nationalgeographic.com/world/0001/mars/index.html

To learn more about the White House Millennium Council Youth Initiative program and the Mars Millennium Project call (310) 274-8787 x150, e-mail them at mars@pvcla.com, or visit the website at http://www.mars2030.net/

You can now view more than 25,000 images of the planet Mars that were taken by NASA's *Mars Global Surveyor* by visiting a website on the Internet. This collection is the largest one-time release of images for any planet in the history of space exploration. Go to: http://www.msss.com/moc_gallery/

ABOUT THE AUTHOR

Tanya Lee Stone is a former editor of children's books who now writes full time. She holds a master's degree in science education and is the author of more than a dozen books for kids, including *Saturn, Rosie O'Donnell: America's Favorite Grownup Kid,* and *The Great Depression and World War II.* She lives in Burlington, Vermont, with her husband, Alan, and her son, Jacob.

INDEX

age, 38
aphelion, 33, 58
astronomers
  ancient, 6
  early, 8–17
  twentieth-century, 18–29, 25, 32, 47, 56
astronomy, 58
atmosphere, 11, 20, 32–33, 53
axis, 11, 30, **31**

basalt, **10**
Brahe, Tyco, 8

cameras, 53
canals, 12
canyons, 42
carbon dioxide, 32, 41
Cassini, Domenico, 11
chlorine, 44
Climate Orbiter, 25
color, 33–37, **35**
concave, 58
convex, 58
Copernicus, Nicolaus, 9
core, 38, **40**
  See also interior
craters, 38
crust, 38, **40**
  See also sub-surface

day, 6, 11, 23, 31, 56
Deimos, 12–13, 34–35
diameter, 56
distance
  from Earth, 17, 29
  from moons, 34
  from Sun, 6, 9, 33, 47, 56
drills, 55
dust, **7**, 27, 33–37, **35**

Earth
  distance from, 17

similarities, 6, 30, 31
elevations, **39**
European Space Agency (ESA), 51–54

fact sheet, 56
fossils, 44
future exploration, 55–57

Galilei, Galileo, 14
Gamma Ray Spectrometer (GRS), 50–51
gamma rays, 51
Global Surveyor, 27–29, **28**, 37, 38, 41–42, 43,
  46, 51
gravity, 37, 56

Hall, Asaph, 12–13, 34
heliocentricity, 9
Hellas Plain, 43
Herschel, William, 11
hot spots, 42
Hubble telescope, **16**, 33
Huygens, Christiaan, 9–10, 16

ice, **32**, 33, 41
images, **7**, **10**, 20–21, **21**, 27, 29, **41**, 43, 46,
  50, **56**
infrared technology, 53
interior, 38, **40**, 53
  See also sub-surface

Kepler, Johannes, **8**, 8–9
Kuiper, Gerard, 33

laboratory, 55
landing vehicles, 26, **27**, 51, 53–55, **54**
life, 11, 12, 17, 20, 21, 23, 44, **45**, 51, 55
lines, 37
Lippershey, Hans, 14
Lowell, Percival, **12**, 12–17

magnetic field, 29
  See also gravity
mantle, 38, **40**

Mariner program, 18–21, **22**, 37
Mars Express mission, 51–55
Mars 1, 18
Mars Radiation Environment Experiment
   (MARIE), 50
Mars Surveyor Orbiter, 48–51, **49**, **51**
meteor, 58
meteorites, 44, **45**, 58
MGS. *See Global Surveyor*
moons, 12–13, 34–35, **36**
mountains. *See* volcanoes

NASA, 27
See also Climate Orbiter; Global Surveyor;
   Mariner program; *Mars Surveyor*
      Orbiter; Observer; Pathfinder; Polar
      Lander; Reconnaissance Orbiter;
      Surveyor missions; *Viking* program
Nirgal Vallis, **47**

observatories, 13–17
Observer, **24**, 25
ocean, 46
Olympus Mons, **21**, 42
Ophir Chasm, **43**
opposition, 58
orbit, 58
   of Mars, 9, 30, 33
   of moons, 34
   of spacecraft, **22**, 29

Pathfinder, **19**, 25–27, **26**
perihelion, 33, 58
Phobos, 12–13, 35, **36**
Phobos program, 25
plains, 43, **43**
planets, motion of, 8–9
polar caps, 6, 11, **32**, **41**, 41–42, 43, 51
Polar Lander, 25

radioactivity, 42, 50
   See also gamma rays
Reconnaissance Orbiter, 54l
revolution, around Sun, 30, 47, 56
riverbeds, 46
rock(s), **10**, 27, **35**, 55
rotation, 11, 23, 31, 56, 58
rovers, 54, 55

See also landing vehicles
Russian programs, 18, 20, 23, 25
   See also European Space Agency

sand dunes, 43
Schiaparelli, Giovanni, **12**, 12–13
seasons, 6, 11, 23, 30
size, 17, 47, 56
   of moons, 34
soil, 33–37, 42, 51
Sojourner, **26**, 26–27
southern hemisphere, 38
Space Age, 18
Stickney Crater, 34
storms, 33–37, **35**
sub-surface, 55
   See also crust; interior
Sun
   as center of universe, 9
   distance from, 6, 9, 33, 47, 56
surface, 27, **35**, 38–43, 46–47, 51, **52**
   of Phobos, **36**
Surveyor missions, 51
Syrtis Major, **10**, 11

telescopes, 14–17, **15**, **16**
Tharsis Bulge, **21**, 38, 42
Thermal Emission Imaging System
(THEMIS), 50

ultraviolet technology, 53

Valles Marineris, **42**, **43**
Viking program, **7**, **22**, 23, **24**, 32, 41
   images, 7
volcanoes, **10**, **21**, 38

water, 11, 20, 27, 44, 46–47, 51, 52
   See also ice
weather, 6, 23, 30, 32–33, 38
websites, 61
weight, on Mars, 37
wind, 23, 33–37, 53

year, 30, 47, 56

Zond program, 20